ETA/Cuisenaire®

SCIENCE

ETA
Cuisenaire

Vernon Hills, Illinois

ETA/Cuisenaire® Elementary Science Dictionary
ETA 55836
ISBN 978-0-7406-5321-6

ETA/Cuisenaire • Vernon Hills, IL 60061-1862
800-445-5985 • www.etacuisenaire.com

Printed in China.

08 09 10 11 12 13 14 15 16   10 9 8 7 6 5 4 3

Introduction

The ETA/Cuisenaire® Elementary Science Dictionary is an essential guide to the scientific language and concepts used in the United States school systems. It has been written with the young reader in mind, giving clear, simple, and concise definitions. Most definitions also include a colorful photo or diagram to assist with understanding. The words covered are those words that students are likely to encounter in their elementary science classes.

This book is a handy, easy-to-use and easy-to-carry reference for all young students. Many everyday words take on a special meaning when we use them in science. Many other words are not everyday words at all. They are unique to scientific language and may be strange to students. Learning these words is an essential part of learning most science concepts. The dictionary will assist students in this task and will also aid parents who are helping their children at home.

Useful reference charts include common abbreviations, measurement conversions, list of elements, and periodic table.

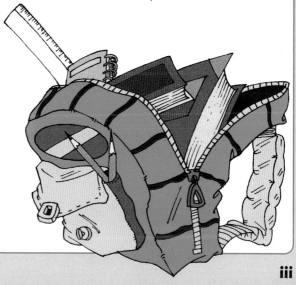

Contents

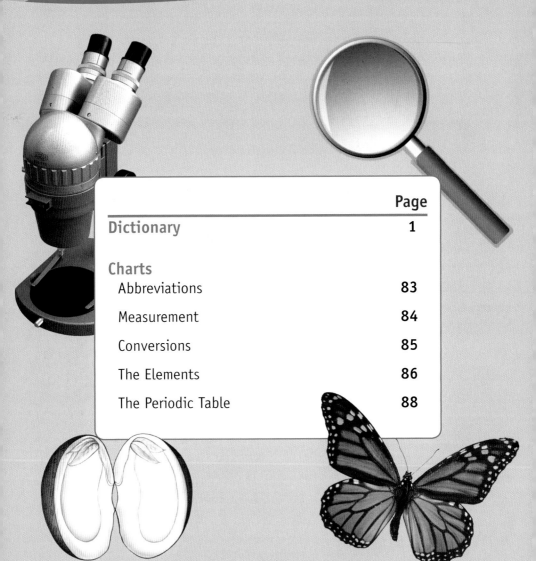

A

absorption
- the taking in of nutrients by the body
- the taking up of light, heat, and other energy by molecules
- the taking up of one substance by another

a sponge picks up water by absorption

acceleration
- rate at which the velocity of an object changes
- acceleration is usualy denoted by the letter *a*
- average acceleration is change in velocity divided by change in time

acid
- a chemical that gives up hydrogen ions (H⁺) when it is dissolved in water
- has a pH below 7

acid rain
- rain that has been made acidic by air pollution
- can cause damage to buildings, statues, forests, and lakes

adaptation
- a structure or behavior that helps an organism survive in its surroundings
- the webbed feet of a duck are an adaptation that helps the duck swim

adolescence
- period between childhood and adulthood
- generally starts at about 14 years for boys and 12 years for girls

adult
- a grownup
- a fully developed organism

A

adulthood
- period after adolescence

air
- a mixture of gases that covers Earth
- made mostly of nitrogen and oxygen

air pressure
- the pressing down of air on Earth
- caused by the weight of the air

algae
- plantlike organisms that have no roots, stems, or leaves

- they contain chlorophyll
- most live in water

amino acids
- a class of nitrogen-containing compounds

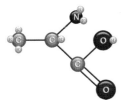

- the building blocks of proteins
- the human body contains 20 amino acids

amphibian
- group of cold-blooded, vertebrate animals
- they can live in water and on land
- they lay their eggs in water

toad   frog

salamander

examples of amphibians are frogs, toads, salamanders, and newts

anemometer
- instrument used to measure wind speed

aquarium
- glass tank that is filled with water and used to keep fish

arteries
- blood vessels that carry blood from the heart to different parts of the body

arthropods
- group of invertebrate animals with hard exoskeletons, jointed legs, and segmented bodies

insect   spider

scorpion

crustacean

examples of arthropods are insects, spiders, scorpions, and crustaceans

asteroid
- a small rocky object that orbits the Sun

- most asteroids orbit between the orbits of Mars and Jupiter

- the largest of these asteroids is called Ceres

astronaut
- person who travels to space

astronomy
- the scientific study of the universe and everything found in it

atmosphere
- air that surrounds Earth

atom
- smallest unit of an element

- contains all the properties of the element

- made up of protons, neutrons, and electrons

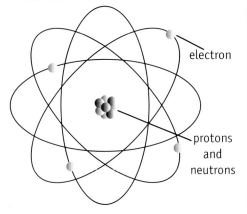

atomic number
- number of protons in the nucleus of an atom

- usually denoted by the letter Z

atomic weight
- the average mass of the atoms of an element

attract
- to pull in or draw toward

- magnets attract objects made of certain materials, such as iron or steel

magnets attracting each other

axis
- imaginary line around which something rotates

- Earth rotates around an axis called the axis of rotation

bacteria
- very small organisms that cannot be seen with the naked eye

- made of only one cell

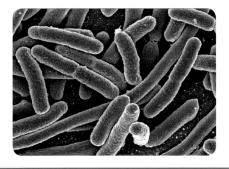

balance
- instrument used to measure the mass of an object

bark
- outermost protective layer on the stems of trees

- made of living cork cells on the inside and dead cells on the outside

barometer
- instrument used to measure the pressure of the atmosphere

battery
- device used to store electricity

beetles
- group of insects that have hard front wings

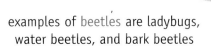

examples of beetles are ladybugs, water beetles, and bark beetles

B

biologist

- scientist who studies living organisms

- conducts research to find out how organisms live and interact with each other

Charles Darwin was a biologist

bird

- a warm-blooded, feathered, vertebrate animal

- birds have wings

- most birds can fly

bladder

- body organ that is used to store urine

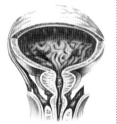

blizzard

- heavy snowfall along with strong wind and extreme cold

blood

- fluid that flows in the entire body

- carries oxygen and nutrients to body tissues

- carries carbon dioxide and other wastes away from tissues

blood vessel

- pipe-like structure through which blood flows

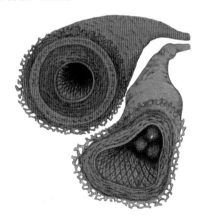

body system
- group of body organs and tissues that work together

boiling point
- temperature at which any liquid starts to boil
- at the boiling point, a liquid starts to change into gas

bone
- hard tissue that forms the skeleton
- bones protect soft organs of the body

bone marrow
- soft tissue found within the cavities of bones
- blood cells are produced in the bone marrow

bone marrow

botany
- the scientific study of plant life

brain
- main part of the nervous system
- made of nerve cells and nerve fibers
- the control center of the body
- located inside the skull

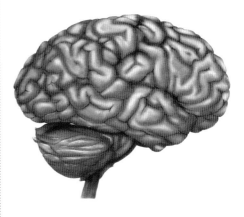

breathing
- act of taking air inside the lungs and letting it out

buoyancy
- force that acts on objects immersed in water or any other liquid

- enables the object to float on water or another liquid

buoyancy makes ships float on water

camouflage

- the color or body shape of an animal that makes the animal hard to see in its surroundings

- helps animals avoid predators or catch food

capillary

- smallest blood vessel, through which substances can pass in and out of the bloodstream

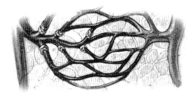

carbohydrate

- a nutrient in food from which the body gets energy

- bread, pasta, and cereal are good sources of carbohydrates

  examples of carbohydrates are sugars, starches, and cellulose

carbon

- an element

- denoted by the symbol C

- has atomic number 6

- the basis of all living matter

carbon dioxide

- a colorless and odorless gas that is made of carbon and oxygen

- denoted by the formula CO_2

carnivore

- animal that eats other animals

- also called meat eaters

cell

- basic structural and functional unit of all organisms

- also called the "building block of life."

- some organisms, such as bacteria, are made of one cell, while humans are made of almost 100,000 billion cells

C

cell membrane

- outer boundary of a cell

- controls which substances enter or exit the cell

cell membrane

cell wall

- rigid layer surrounding the cells of plants

- gives form, shape, and protection to cells

- made of cellulose

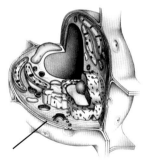

cell wall

Celsius

- a temperature scale

- abbreviation for Celsius is C

the Celsius scale was named after the Swedish astronomer Anders Celsius

chemical bond

- force that holds atoms of elements together in a chemical compound

chemical change

- change in the chemical composition of a substance, which produces a new substance with new properties

burning of wood produces ash and smoke, which have different properties than wood

chemical reaction

- process in which two or more chemical substances react to yield substances of different composition and properties

 + =

sodium and chlorine react together to produce sodium chloride or common salt

chemical weathering

- process by which rocks and minerals are transformed into new chemical compounds

the change of feldspars into mud is an example of chemical weathering

10

child

- young human who has not reached adolescence

chlorophyll

- green pigment found in plants

- captures energy from sunlight

chloroplast

- organelle found in plant cells that contains chlorophyll

- carries out photosynthesis

chromosome

- threadlike structure found in the nucleus of a cell

- contains all genetic information

- normal human cell contains 23 pairs of chromosomes

chrysalis

- pupa of a butterfly or moth

circuit

- path through which electric current flows

circulatory system

- body system related to the circulation of blood through the body

- major parts of the circulatory system are the heart and blood vessels

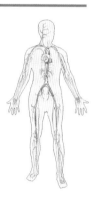

cirrus

- type of cloud

- thin cloud composed of ice crystals

- usually form above 5.5 km

c

classification
- arrangement of organisms into classes, groups, or categories of the same type

climate
- weather conditions over a long period of time in a particular location

cloud
- mass of condensed water droplets or ice crystals that may bring rain
- suspended in the atmosphere above Earth's surface

cold-blooded
- having a body temperature that is controlled by the temperature of the surroundings

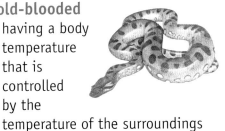

- examples are snakes and fish

cold front
- the front edge of a cold air mass moving in under a warm air mass

colony
- group of organisms of the same species living together

colony of bees

comet
- astronomical object largely made of ice that travels around the Sun in an elliptical orbit

12

communication

- exchange of information between two or more people or groups using symbols, signs, behavior, speech, writing, or signals

community

- all the organisms living together in an area

- organisms in a community interact and depend on one another for existence

compass

- navigational instrument used for finding directions

- made of a magnetic needle that always points north

compound

- chemical substance formed from two or more elements

  water is a compound made from two hydrogen atoms and one oxygen atom (H_2O)

computer

- electronic device used to store and process information

- runs with the help of software

condensation

- process of changing from gas to liquid or solid

conduction

- transfer of heat or electricity from one point to another in a medium

conductor

- substance through which electric current or heat can pass easily

- most metals are good conductors of heat and electricity

coniferous tree

- tree with cones and needle-shaped leaves
- its seeds are produced in cones

examples of coniferous trees are pines, redwoods, and spruces

conservation

- the practice of protecting resources from loss or waste

constellation

- group of stars that forms a familiar shape or pattern in the sky

- there are 88 constellations

consumer

- part of a food chain that feeds on other living organisms

- organisms that feed on green plants are primary consumers, while those who eat other animals are secondary and tertiary consumers

continents

- the seven main land masses on Earth

- Africa, Antarctica, Asia, Australia, Europe, North America, and South America

control

- the part of an experiment that other parts are compared with

convection
- movement of heat from one point to another within a gas or liquid

core
- the innermost part of Earth

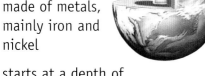

- made of metals, mainly iron and nickel

- starts at a depth of approximately 2,900 km

crab
- group of generally aquatic crustacean animals

crater
- a bowl-shaped hole or depression in the ground

- formed by volcanic eruptions or by the impact of an asteroid or meteorite

crust
- outer layer of Earth

- about 10 km thick under the oceans and up to about 50 km thick on the continents

crystal
- a piece of a mineral or other chemical in which atoms are arranged in a regular pattern

culture
- a growth of microorganisms in a laboratory for research

cumulus
- type of cloud

- dense cloud with sharp outlines, and taking forms of rising mounds, domes, or towers

- formed in the troposphere below about 2,400 m

15

current

- flow of electric charge through a conductor

- electric current is measured in amperes and is usually denoted by the symbol 'i'

- movement of water in an ocean or river

cycle

- a process that happens over and over

- an example is the water cycle

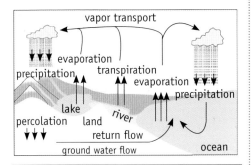

cytoplasm

- jellylike substance in a cell, between the nucleus and the cell membrane

data

- facts collected in an organized manner

- data can be communicated, interpreted, or processed

decay

- decomposition or disintegration of organisms or other substances into simpler forms of matter

- can happen by the action of fungi or bacteria

deciduous tree

- tree that sheds its leaves annually

oak, hickory, and ash are examples of
deciduous trees

decomposer

- organism that breaks down organic matter into substances that can be used by other organisms

- fungi and bacteria are decomposers

degree

- unit for measuring temperature

- unit for measuring angles

- denoted by °

delta

- triangular, fan-shaped deposition of soil at the mouth of a river

- formed as a river carries sediment down stream

17

density
- mass per unit of volume

deposition
- geological process whereby material is added to a landform by wind, water, or ice
- also known as sedimentation

desert
- dry, barren, often sand-covered land area, where the precipitation is less than 25 cm per year
- usually characterized by hot daytime temperatures

dew
- small droplets of water that form on objects close to the ground in the morning or evening
- formed by condensation of water vapor

diamond
- a crystalline form of carbon
- a precious transparent gemstone

- the hardest mineral
- formed due to high compression in the earth over millions of years

diaphragm
- thin muscle below the lungs that is used for breathing
- separates the chest from the abdomen

diffusion
- intermixing of atoms and molecules in solids, liquids, and gases

- the movement of suspended or dissolved particles from an area of high concentration to an area of low concentration

  a drop of colored water mixing with clear water is due to diffusion

digestion

- process by which the body breaks down food into simple substances and uses them for energy, growth, and cell repair

- proteins are broken down to amino acids, starch into glucose, and fats to glycerol and fatty acids

digestive system

- body system that breaks down food into chemical components that the body can absorb and use

- the main organs of the digestive system are mouth, stomach, liver, and small and large intestines

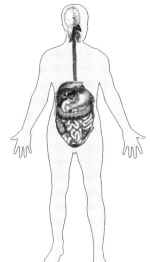

dinosaur

- extinct carnivorous or herbivorous reptiles that existed during the Mesozoic era between 248 and 65 million years ago

discovery

- the learning or uncovering of information

- usually the uncovering of something in nature

disease

- abnormal condition of the body usually caused by germs or other environmental factors

dissolve

- to spread out and disappear into another substance

19

distance
- length between two places

DNA
- stands for deoxyribonucleic acid

- a chemical that carries genetic information about an organism

- is found inside the nucleus of a cell

- consists of two strands made of adenine (a), thymine (t), guanine (g), and cytosine (c)

drag
- force that acts against forward motion

- slows a moving object down

drag acts on flying airplanes

drought
- long period of dry weather or shortage of rainfall

- causes damage to plants

drug
- chemical substance that affects processes in the body, used mainly for treating or preventing disease

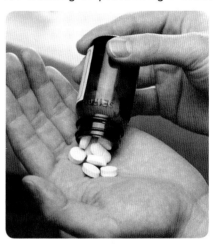

ear

- structure in some animals that detects sound

- fruiting body of a grain plant, such as corn or wheat

earthquake

- sudden shaking and vibration of the ground

- caused by the release of stress in the crust and most commonly related to movement of Earth's tectonic plates

eclipse

- event in which one celestial object blocks the view of another

in a solar eclipse, the Moon blocks our view of the Sun

ecosystem

- all the living and nonliving things in an area

rainforests, deserts, coral reefs, and grasslands are examples of ecosystems

effort

- the amount of input force required to do work by using a simple machine

egg

- first stage of life for many animals, such as birds

- consists of an ovum or embryo together with food reserves

electricity
- form of energy associated with charged particles, mainly electrons

electromagnetism
- magnetism produced by an electric current

electron
- a basic particle in an atom
- orbits around the nucleus of an atom
- carries a negative charge

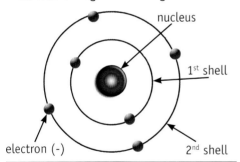

nucleus

1st shell

electron (-)

2nd shell

electronics
- the study and use of electric circuits

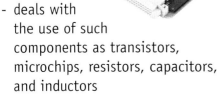

- deals with the use of such components as transistors, microchips, resistors, capacitors, and inductors

element
- substance that cannot be separated into simpler substances by chemical processes
- composed of a single kind of atom
- elements can combine to form compounds

embryo
- early stage of the development of an animal or plant
- in animals, the embryo forms in the uterus or egg
- in plants, the embryo forms in the seed

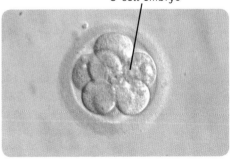

8-cell embryo

endangered species
- organisms that are in danger of extinction

endocrine system
- a body system of glands that secrete hormones into the blood

energy
- ability or capacity to do work

- common forms of energy include thermal, mechanical, electrical, and nuclear energy

engine
- machine that converts any form of energy into mechanical power

a railroad locomotive is an engine used to move trains along railway tracks

environment
- all of an organism's surroundings

enzyme
- protein found inside living organisms

- increases the rate of chemical reactions inside the body

epicenter
- the area on Earth's surface that is directly above the point where an earthquake originates

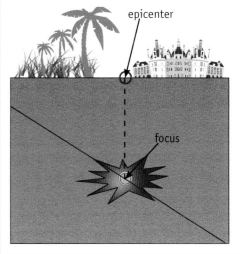

equation
- an expression of a chemical reaction using symbols

$$2H_2 + O_2 \longrightarrow 2H_2O$$

- an expression showing that two quantities are equal

$$8 \times 2 = 12 + 4$$

equator

- imaginary line that divides Earth into two halves and is the same distance from the North and South Poles

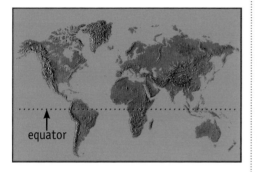

equator

erosion

- the wearing away of sediment or soil by ice, wind, and water

esophagus

- muscular tube that leads from the throat to the stomach
- used for the passage of food from the mouth to the stomach

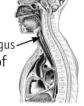

esophagus

evaporation

- process of changing from liquid to gas

evolution

- change in the traits and characteristics of living organisms over millions of years

exhale

- act of breathing out

experiment

- organized procedure to support or disprove a hypothesis

extinct

- no longer in existence

eyes

- organ used to provide vision

Fahrenheit
- a temperature scale
- abbreviation for Fahrenheit is F

fats
- soft, greasy organic compounds found in animals and plants
- made of glycerol and fatty acids, they are insoluble in water

fault
- crack in the Earth's crust
- caused by the movement of the Earth's plates

feather
- the distinctive outer covering on birds

femur
- thigh bone that extends from the pelvis to knee
- longest bone in the human body
- third leg segment of the leg of an insect

ferns
- group of plants that reproduce using spores
- they have no seeds or flowers

fertilization
- union of sperm and egg that forms the embryo in sexual reproduction

fin
- body structure used for swimming and balancing in water
- found on fishes and some other aquatic animals

F f

fish

- cold-blooded, vertebrate animals that live in water, have fins, and breathe through gills

flammable

- capable of catching fire easily

float

- to stay on a liquid surface and not sink down

flood

- overflowing of a large amount of water beyond the artificial or natural boundaries of a stream, river, or other body of water

flower

- reproductive organ of plants

- flower-bearing plants are known as angiosperms

- petals, sepals, stamens, and pistil are the major parts of a flower

food chain

- sequence of organisms that shows what eats what

- each organism uses lower member of the sequence as a food source

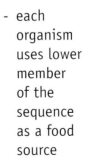

food web

- network of interconnected food chains and feeding relationships in an ecosystem

force

- a push or a pull acting upon an object

forecast

- prediction of future conditions by analysis of available data

- used to predict natural phenomena like weather conditions

forest

- an ecosystem that includes a thick growth of trees

fossil

- remains, traces, or imprints of plants and animals that existed in the past

- many fossils are preserved in sedimentary rocks

fossil fuels

- sources of energy

- formed in the ground over millions of years from the remains of dead plants and animals

> Oil, natural gas, and coal are examples of fossil fuels

freezing point

- temperature at which a liquid turns into solid

> at 0° C, water freezes into ice

frequency

- number of wave cycles per unit time

- usually measured in cycles per second, or hertz

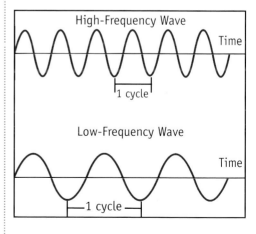

27

friction

- force that resists the motion of an object

- acts between two objects in contact with each other

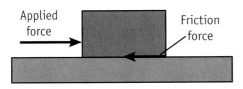

Applied force → | ← Friction force

front

- boundary between two air masses of different density, temperature, and humidity

fruit

- ripened ovary of a flowering plant that contains one or more seeds

fuel

- substance that is burned to produce energy

fulcrum

- fixed point around which a lever moves

Force

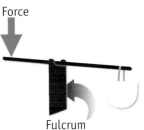

Fulcrum

fungi

- group of plantlike organisms that have no chlorophyll, leaves, flowers, or roots

- they obtain nutrients from dead or living organic matter

molds, mushrooms, and yeast are examples of fungi

fur

- the coat of hair that covers the bodies of many mammals

G

galaxy

- collection of stars, gas, and dust held together by gravitational force

- a galaxy may contain billions or trillions of stars

gall bladder

- pear-shaped sac in the human body used to store bile

- it is attached to the liver

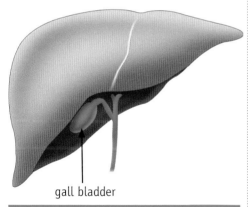

gall bladder

gas

- one of the phases of matter in which molecules move around freely

- does not have fixed volume or fixed shape

gasoline

- liquid fuel produced from refining crude oil

- used to power automobiles

gene

- basic unit of genetics that is composed of DNA

generator

- machine that converts mechanical energy into electrical energy

genetics

- scientific study of how traits pass from one generation to the next

geology

- branch of science dealing with the composition, structure, physical properties, and history of Earth

geotropism
- downward growth of the roots of plants due to the effect of gravity

germination
- first stage of the development of a plant from a seed

germ
- microorganism that can cause disease

- some bacteria and viruses are germs

gills
- respiratory organ found in fish and some other marine animals

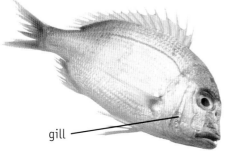

gill

glacier
- large river of ice that is formed over a large period of time through recycling of snow

- moves very slowly due to its own weight

gland
- organ that produces and releases hormones, juices, and enzymes

graphite
- soft form of dark gray or black carbon that is used as pencil lead and lubricant

grassland
- biome in which grass is the primary natural vegetation

- examples include savannas, pampas, campos, plains, steppes, and prairies

gravitational force
- force of attraction between two objects with mass

- $F = G \times \dfrac{m_1 \times m_2}{d^2}$

  where

  F is gravitational force

  m_1 is mass of one object

  m_2 is mass of the other object

  d is distance between the two objects

  G is gravitational constant

gravity
- force that tends to pull objects toward Earth

- force that tends to pull any two objects together

growth
- increase in size of an organism over a period of time

habitat
- natural home of plants or animals

hail
- precipitation in the form of chunks of ice that fall from cumulonimbus clouds
- associated with thunderstroms

hearing
- the ability to detect sound
- one of the five senses

heart
- hollow, muscular organ that pumps blood throughout the body
- humans have a four-chambered heart

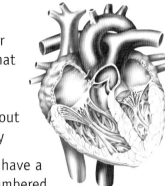

heart rate
- number of times the heart beats in one minute
- normal adult human heart beats about 70 beats per minute

heat
- transfer of thermal energy from one object at a higher temperature to another object at a lower temperature

hemoglobin
- compound made of protein and iron that is found in red blood cells
- carries oxygen to cells and carbon dioxide away from the cells
- gives red color to blood

herbivore
- animal that eats only plant materials

heredity
- passing of certain genetic characteristics and traits from parents to their offspring

hertz
- unit to measure frequency
- cycles per second
- denoted by Hz

hibernate
- to go into a sleep-like condition of partial or total inactivity
- some animals do it to conserve energy during the winter

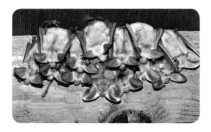

hormone
- chemical substance secreted by glands into the blood and transported to specific organs to regulate particular functions
- essential for metabolism and growth of the body

horsepower
- unit to measure mechanical power or work done over time

host
- an organism on or in which a parasite, or mutual partner lives for nourishment and support

humidity
- amount or concentration of water vapor in a given volume of air

humus
- decomposed residue of plant and animal tissues

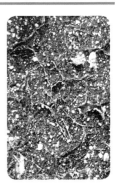

- provides nutrients for plants and increases the ability of soil to retain water

H

hurricane
- tropical cyclone with heavy rainfall and high-velocity wind

hydroelectric
- production of electricity through the use of moving water

Hydroelectric power plant

hydrogen
- the lightest element

- colorless, odorless, highly flammable gas

- denoted by the symbol H

- atomic number is 1

hypothesis
- proposed or tentative explanation of certain scientific facts or observations

igneous rock
- type of rock
- formed from cooling and solidification of magma either on or within Earth's crust

image
- visual representation of an object or scene

immune
- protected from or resistant to a specific disease

inertia
- tendency of an object to remain in the same state either at rest or in motion

infant
- child less than 12 months old

inherited
- genetic characteristics obtained from parents

inorganic
- not originating from living matter
- compounds not based on carbon

input
- something put into a process or action

insect
- largest group of animals, having over 800,000 species
- generally have three body parts, six legs, single pair of antennae, wings, compound eyes, and a hard external skeleton

grasshoppers, beetles, butterflies, moths, ants, bees, and wasps are examples of insects

instinct
- inherited tendency to respond to given situations

insulator
- object or material that resists the flow of electric charge, heat, or sound

interaction
- communication or other contact between individuals or organisms

International Space Station
- research station located in space approximately 360 km above Earth
- joint project of six space agencies

Internet
- worldwide communications network that connects local computers to the computers of academic institutions, research institutes, private companies, and government agencies

intestine
- long, tube-shaped body organ used for digestion and absorption of food

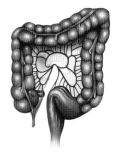

invention
- creation of a new device, method, or process

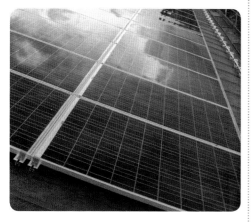

invertebrate
- animal without a backbone

insects, worms, snails, and mussels
are examples of invertebrates

investigation
- systematic study or search for facts

iron
- a common, widely used metal
- the main component of steel
- one of the elements
- denoted by the symbol Fe
- has atomic number 26

J

jaw

- two opposable structures made of bone or cartilage that frame the mouth and hold the teeth

- part of the skull

- used for chewing of food

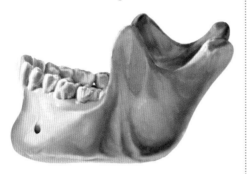

jellyfish

- saucer-shaped free-swimming sea animals that belong to phylum Cnidaria

- have soft jellylike body and stinging tentacles

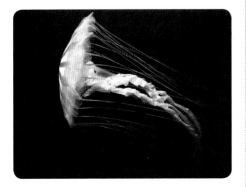

joint

- where two bones are joined together

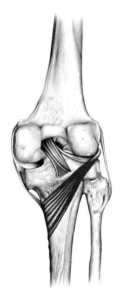

joule

- unit to measure energy

- denoted by the symbol J

kelvin

- unit to measure absolute temperature

- absolute zero on the Kelvin scale is -273° on the Celsius scale

keyboard

- input device for a computer

- consists of various letter, number, and special function keys

kidney

- bean-shaped body organ

- filters wastes from the blood and regulates blood pressure

- removes waste products and excess water as urine

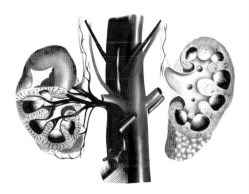

kinetic energy

- energy of a moving object

- denoted by KE

- $KE = \frac{1}{2} \times m \times v^2$, where m = mass and v = velocity

kingdom

- the highest level in the scientific system of classification

- organisms are divided into five kingdoms: Animalia, Plantae, Fungi, Protista, and Monera

L

laboratory
- place for conducting scientific research and experiments

landform
- physical feature on Earth's surface, such as a hill, valley, or canyon

- has distinct characteristics, a recognizable shape, and is produced by natural processes

larva
- early stage of the life cycle of an insect

- stage between egg and pupa

- a caterpillar is the larva stage of a butterfly or moth

latitude
- how far north or south of the equator

- measured in degrees and minutes

latitude

lava
- molten rock that reaches the surface of the Earth through volcanic eruptions

law
- universal principle that describes scientific facts or events in nature

lens
- transparent object made with one or two curved surfaces to bend or focus light

- the part of the eye that focuses images

lever

- simple machine consisting of a rigid bar

- works by moving around a fulcrum or fixed point

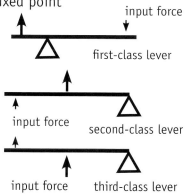

input force

first-class lever

input force

second-class lever

input force

third-class lever

life cycle

- developmental stages of an organism's life

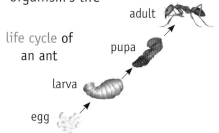

life cycle of an ant

adult

pupa

larva

egg

lift

- upward force that acts on an aircraft and enables it to take off and fly

- generated by the airflow around a wing

ligament

- tissue that connects bone to bone or cartilage to bone

- supports a joint

light

- type of energy or electromagnetic radiation with a wavelength that is visible to the eye

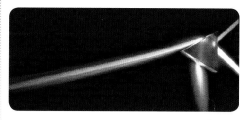

light year

- distance traveled by light in one year

- equal to 9.4607×10^{12} km

lightning

- visible electric discharge from cloud to cloud or from cloud to Earth

- accompanied by the emission of light and sound

L

liquid

- a phase of matter with loosely connected molecules

- has a definite volume but its shape is determined by the container it fills

liter

- a unit of volume in the metric system

liver

- largest internal organ in the body

- has a number of functions, such as metabolism of protein, carbohydrates, and fats; glycogen storage; protein synthesis; and bile secretion

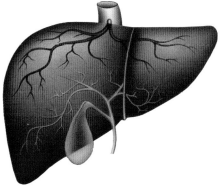

living thing

- anything that has life, is alive

load

- an object that is moved by a simple machine

loam

- type of soil composed of sand, silt, clay, and organic matter

- best soil for the growth of most plants

longitude

- how far east or west of the prime meridian at Greenwich

- measured in degrees and minutes

longitude

lubrication

- reducing friction between two moving surfaces by applying grease or other coating

lunar eclipse

- eclipse of the Moon

- occurs when Earth comes in a direct line between the Sun and the Moon

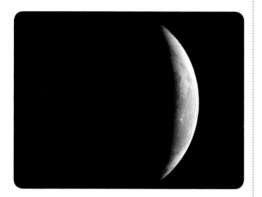

lung

- body organ that belongs to the respiratory system

- supplies oxygen to the body and removes carbon dioxide from the body

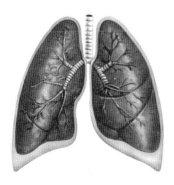

luster

- brightness of a mineral

mach
- a number that indicates the speed of an object relative to the speed of sound

machine
- device that does work

magma
- molten rock beneath the surface of the Earth

magnet
- object that has a magnetic field and attracts iron and steel

magnetic field
- region of force lines around a magnet

mammal
- warm-blooded vertebrate that has mammary glands for producing milk to nourish the young and has hair on its skin

mantle
- layer of Earth below the crust and above the core

- largest layer of Earth composed of very hot, dense, flowing rock

marsh
- area of soft, wet, or periodically submerged land

- a wetland ecosystem with tall grasses

mass
- quantity of matter in an object
- mass is usually expressed in kilograms

matter
- anything that has mass and occupies space
- matter can exist in four states: solid, liquid, gas, and plasma

measurement
- determining the size, quality, or quantity of an object

- system of measuring based on a systematic application of a standardized procedure

mechanical
- related to machines or the movement of structures and objects

mechanical weathering
- weathering of rocks by physical processes without changing their chemical composition

melting
- process of changing a solid to a liquid by heating

melting point
- the temperature at which the state of a substance changes from solid to liquid

metal
- mineral element that conducts heat and electricity

examples of metals are iron, copper, silver, and lead

metamorphic rock
- rock that has undergone chemical or physical changes as a result of extreme heat, pressure, or chemically active fluids

metamorphosis

- change in the form, shape and structure of an object

- transformation of an animal during its life cycle from a larval state to an adult state

   transformation of a tadpole to a frog is an example of metamorphosis

meteor

- mass of rock or metal that has entered Earth's atmosphere

- commonly called a shooting star or a falling star

meteorologist

- scientist who makes a scientific study of weather and predicts the weather

metric system

- universally used system of measurements in which all units are based on multiples of 10

- system of units based on the meter as the unit of length, the kilogram as the unit of mass, and the second as the unit of time

microorganism

- microscopic living organism

- microorganisms include bacteria, fungi, algae, and protozoa

microscope

- an instrument for viewing objects that are too small to be seen by the naked eye

- optical instrument that can produce a magnified image

migration

- movement of living beings from one biome to another

- seasonal or annual movement of animals, fish, and birds in search of food or shelter

Milky Way

- a spiral galaxy containing our solar system

- bright band of stars stretching across the night sky

mineral

- natural, usually inorganic compound having a definite chemical composition and crystal structure

- building blocks of rocks

mirror

- reflective surface that is smooth enough to form an image

mitosis

- process of cell division

- division of the nucleus in cells to form two new nuclei containing the same number of chromosomes as the parent nucleus

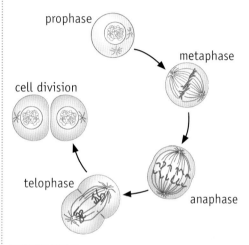

prophase

metaphase

cell division

telophase

anaphase

mixture

- two or more substances mixed together in such a way that each remains unchanged

- combination of substances which can be separated by physical techniques such as filtering

47

M

Mohs' scale
- scale used to indicate relative hardness of rocks and minerals
- uses numbers 1 to 10, where 1 means soft as talc and 10 means hard as diamond

Mohs' scale was invented by Friedrich Mohs

moisture
- water or other liquid causing wetness

molecule
- smallest unit of two or more atoms that retains all the physical and chemical properties of a particular substance

momentum
- the tendency of an object to remain in motion
- it is the product of the mass and the velocity of an object

monsoon
- seasonal wind, especially in the Indian ocean and southern Asia
- reversal in wind direction that causes a shift from a dry season to a rainy season or from a rainy season to a dry season

moon
- natural rocky body that orbits around a planet
- Earth's only natural satellite

moon phases
- changing face of the Moon as seen from Earth
- the phases occur through a cycle that repeats every four weeks

motion
- movement of an object in relation to its surroundings

motor
- device that converts electrical energy into mechanical energy

mountain
- landmass that extends above its surroundings
- a raised part of Earth's surface

mouse
- a handheld pointing device for computers

- used to make selections and to position the cursor on the screen

movement
- process of an object changing position

muscle
- body tissue that functions as a source of power

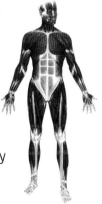

- made of fibers organized into bands or bundles that cause bodily movement by contraction

music
- rhythmic arrangement of sounds and pitches
- sound that is pleasing to the ear

negative charge
- electrical charge created by having more electrons than protons

nephron
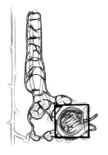
- basic structural and functional unit of the kidney

- filtering unit of the kidney, which maintains the body's chemical balance

nerve
- building block of the nervous system

- bundle of fibers that transmits electrical messages

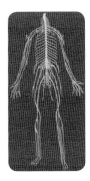

neuron
- a nerve cell

neutron
- uncharged particle in the nucleus of an atom

newton
- a unit of force

niche
- the function of an organism in its habitat

- set of ecological conditions that a living thing requires for its survival and successful reproduction

nitrogen
- odorless, colorless gas

- an element denoted by N

- has atomic number 7

- the main component of air

- found in all living things

nonliving things
- things that are not alive

soil, rocks, and water are examples of nonliving things

nonmetal
- element that is not a metal

nonrenewable resource
- a resource that cannot be replaced once it is used up

minerals, oil, gas, and coal are examples of nonrenewable resources

nose
- body organ used to smell and breathe

nuclear fuel
- material that can be used to generate nuclear power

nucleus
- central, positively charged region of an atom
- central part of a cell responsible for growth and reproduction
- contains all the DNA and RNA

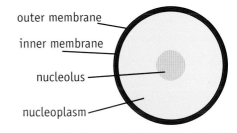

outer membrane
inner membrane
nucleolus
nucleoplasm

nutrient
- substance that provides nourishment for an organism
- any element or compound necessary for an organism's metabolism, growth, or other functioning

nutrition
- practice of taking in foods for growth, energy, and health
- branch of biology concerned with the study of nutrients

nymph
- intermediate stage of the life cycle of some insects that looks like a small adult with no wings

observe
- make a visual examination

- pay special attention to details or behavior and arrive at a conclusion

ocean
- any of the four major bodies of salt water on Earth: Arctic, Atlantic, Indian, and Pacific Oceans

offspring
- the children or young of a parent or parents

- product of reproduction

omnivore
- organism that eats both plants and animals

opaque
- able to block the passage of light

brick and wood are opaque materials

orbit
- path followed by a body when it revolves around another body

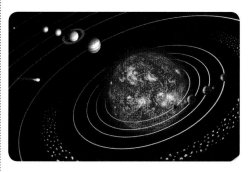

ore
- mineral deposit containing useful metals, such as iron

organ
- group of tissues that perform a specific function or a set of functions

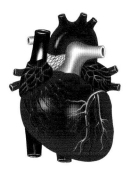

heart, kidney, leaf, and stem are examples of organs

organic matter
- carbon-based substances derived from living things

organism
- individual living system that is capable of reproduction, growth, and response to stimulus

osmosis
- tendency of a fluid to pass through a permeable membrane into a less concentrated solution to equalize the concentrations on both sides of the membrane

output
- something generated by a process or action processed and displayed either in the form of a hard copy or digital format

ovary
- pair of female reproductive glands that produces egg cells in animals

- the part of a flower in which seeds form

ovary

ovule
- part of a flower that can become a seed if it is fertilized

oxygen
- odorless, colorless gas

- the second most abundant component of air

- an element denoted by O

- has atomic number 8

ozone layer
- region of the atmosphere from 19 to 48 km above Earth's surface

- layer of gaseous ozone (O_3) in the stratosphere that shields Earth from much of the Sun's ultraviolet rays

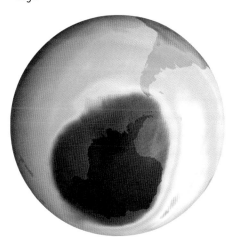

pancreas

- gland behind the lower part of the stomach

- secretes digestive enzymes and the hormone insulin into the blood

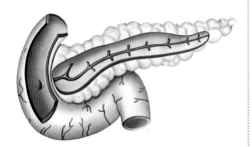

Pangaea

- supercontinent that was the only landmass on Earth about 250 million years ago

parasite

- organism that grows, feeds, and lives on or in an organism of another species

periodic table

- systematic chart that lists the elements in order of increasing atomic number and by their electron arrangements

pesticide

- chemical used to kill pests, such as weeds and insects

phase of matter

- physically distinct form of matter having uniform composition and properties

  solids, liquids, and gases are phases of matter

phloem

- plant tissue that transports food and nutrients from the leaves to the other parts of the plant

photosynthesis

- process used by plants to make food

- occurs in cells in the leaves that contain chlorophyll

- plants make food by using water, carbon dioxide, and sunlight

phylum

- second highest level in the scientific classification of organisms

physical change

- change that does not involve the generation of any new substances

pistil

- female reproductive organ of a flower

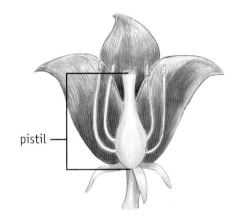

pistil

pitch

- property of a sound, especially a musical tone

- related to the highness or lowness of a sound

planet

- large celestial body orbiting around the Sun

- our solar system has eight planets: Mercury, Venus, Earth, Mars, Jupiter, Saturn, Uranus, and Neptune

Mercury   Venus   Earth   Mars

Jupiter   Saturn   Uranus   Neptune

plastic

- type of man-made materials composed of large organic molecules

- able to be shaped and formed using pressure and heat

plate tectonics

- theory that states that Earth's surface is made of large plates that remain in continuous movement, thus changing the position of continents and oceans

pod

- term used to denote a group of whales

- an elongated seed case common to bean plants

poisons

- substances that are harmful to living tissues and can cause death to organisms

polar regions

- areas surrounding Earth's poles, north of the Arctic circle and south of the Antarctic circle

- characterized by extremely cold temperatures, with 24-hour daylight in summer and 24-hour darkness in winter

poles

- extreme points at the ends of the north and south parts of Earth

pollen

pollen

- powder produced by anthers, the male part of a flower

- can fertilize the ovule on the female part of a flower

pollination

- transfer of pollen from the male part of a flower to the female part

- pollens are carried by insects, birds, animals, rain, and wind

pollution

- release of harmful substances into water, soil, or air

population

- number of people occupying a defined space

- all members of a species living together in an area

pores

- openings in plant leaves that are used for respiration and photosynthesis

- also known as stoma

- openings in the skin used for sweating and secretion of oils

- spaces between grains of soil

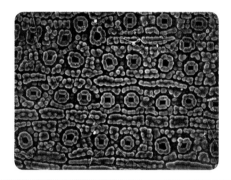

positive charge

- electrical charge that is created by having fewer electrons than protons

potential energy

- stored energy

- for example: the energy stored in a compressed spring

power

- amount of work done per unit of time

power station

- place for the generation of electric power

prairie

- North American grassland characterized by a variety of grasses and wildflowers with few or no trees

precipitation

- water released from the air in the form of snow, rain, sleet, freezing rain, or hail

predator

- an animal that hunts and kills other organisms for food

pressure

- application of force in an area of surface

- denoted by p

- $p = F/A$, where p is the pressure, F is the force, A is the area

prey

- an animal killed and eaten by a predator

primary colors

- three colors of light that cannot be made by mixing other colors: red, green, and blue

- all other colors of light can be made by mixing these colors

- also, the basic colors of paint that are mixed to make other colors: red, blue, and yellow

- also, the basic colors of ink that are used in printing: cyan, magenta, and yellow

prism

- three-dimensional triangular-shaped glass or other transparent material

- used to refract light, reflect it, or break it up into its constituent colors

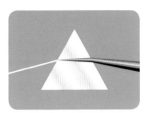

processor

- computer device that processes received information and controls the execution of program instructions

producer

- organism capable of producing its own food, usually through photosynthesis

product

- substance that is obtained from a chemical reaction

projectile

- object that is launched into the air by the application of a temporary force

protein

- large, complex compound found in all animal and vegetable tissues

- made up of one or more amino acids

- obtained in diet from meat, fish, eggs, milk, and vegetables

protist

- single-celled organism belonging to the kingdom protista

proton

- one of the two basic particles in an atom's nucleus

- has a positive electrical charge

protozoan

- single-celled organism that belongs to kingdom protista

- some act as parasites and can cause many diseases, such as malaria and dysentery

pull

- force that causes motion toward its source

pulley

- a simple machine made of a wheel with a groove along its edge

- used to reduce the amount of force needed to lift a heavy weight or to change the direction of that force

weight

force

pupa

- third developmental stage in the life cycle of an insect

- stage between larva and adult

push

- force that causes motion away from its source

59

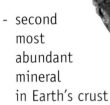

quartz

- mineral made of silicon and oxygen

- second most abundant mineral in Earth's crust

quill

- long, strong feather of the wing or tail of a bird

- writing pen made from a feather

R

radar

- system of detecting and locating a moving object

- short for Radio Detection And Ranging

radiation

- energy transmitted in the form of electromagnetic waves

radio wave

- electromagnetic wave used for long-distance communication

- has a wavelength between 0.5 cm and 30,000 m.

rain

- precipitation that falls in the form of liquid water

rain gauge

- instrument used to measure the amount of rainfall and other forms of precipitation

rainforest

- dense, relatively warm, forest with a high annual rainfall found in tropical areas

ramp

- inclined pathway connecting different levels

- sloping runway

red blood cell

- blood cell containing hemoglobin that carries oxygen to all parts of the body

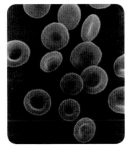

- also called an erythrocyte

reactant

- chemical substance that undergoes changes during a chemical reaction

61

R

recycling
- reuse of materials that would otherwise be considered waste

- process by which materials are converted into new products

reflection
- bouncing of light off of a surface

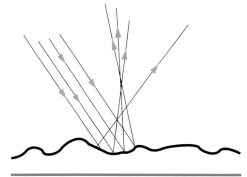

reflex
- automatic, involuntary response of the nervous system to a stimulus

refraction
- bending of light as it passes from one medium to another

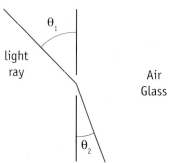

θ_1

light ray

Air
Glass

θ_2

renewable resource
- natural resource that has the capacity to replenish itself

  biomass and solar energy are examples of renewable resources

repel
- to cause an object to move away

reproduction
- process by which a plant or animal produces one or more new individuals similar to itself

reptile
- air-breathing, cold-blooded vertebrate that has a tough covering of plates and/or scaly skin

- lays its eggs on land

turtles, crocodiles, snakes, lizards are examples of reptiles

research
- systematic investigation to discover or revise facts or theories

- to gather information

resistance
- degree to which a body opposes the passage of an electric current

- denoted by R

- R = V/I, where V = voltage and I = current

respiratory system
- body system that engages in gas exchange

- system that takes in oxygen and expels carbon dioxide

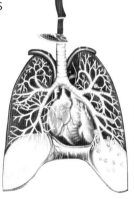

retina
- thin layer of cells at the back of the eyeball that converts light into nervous signals

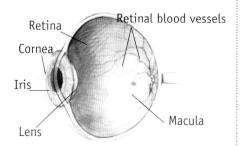

Retina
Cornea
Iris
Lens
Retinal blood vessels
Macula

revolution
- orbital movement of one body around another

rib cage
- bony structure formed by the ribs that surrounds and protects the lungs, heart, and other internal organs of the body

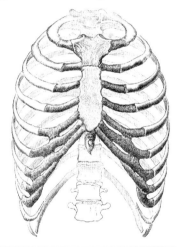

river
- a natural body of flowing water larger than a stream

rock

- naturally occurring solid mixture of minerals

rocket

- vehicle that moves at high speed and is propelled by ejection of fast-moving exhaust gas

root

- part of a plant that usually lies below the surface of the soil

- holds the plant in position and draws water and nourishment from the soil

rotation

- spinning of a body around its own axis

saliva
- fluid produced in the mouth by salivary glands
- moistens food and begins the process of digestion.

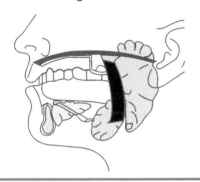

salt
- chemical compound formed by the reaction of an acid and a base

satellite
- object that orbits around a planet
- can be natural such as the Moon or man-made like a space station

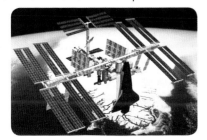

scale
- instrument used to measure weight

scientific method
- procedure used by scientists to test hypotheses

scientists
- person who studies science

screw
- one of the simple machines
- made of a shaft with an inclined plane wrapped around it
- can be used as a fastener

season
- one of the major divisions of the year based on broad climatic patterns

a year is generally divided into four seasons: spring, summer, autumn, and winter

65

S

sediment
- weathered rock particles, usually deposited in layers

sedimentary rock
- type of rock formed by sedimentation

sandstone is an example of a sedimentary rock

seed
- mature reproductive embryo of a flowering plant covered with a protective coat and ready to grow into a new plant

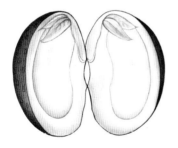

sense
- perceive by physical sensation

shadow
- dark area that is produced when light is blocked by an object

simple machine
- elementary device that works with the application of a single force

- makes work easier

- lever, wheel and axle, pulley, inclined plane, wedge, and the screw are the six simple machines

sink
- go to the bottom of a fluid, not float on its surface

skeleton
- rigid framework of bones

- gives structure and support to the body and protects soft organs and tissues

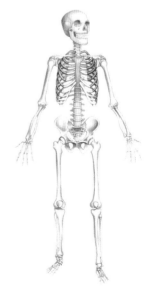

skin

- largest organ of the body

- outer covering of the body used for insulation, vitamin D production, sensation, and excretion

skull

- skeleton of the head that protects the brain

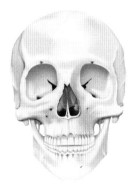

sleet

- precipitation in the form of frozen or partially frozen rain

smell

- sensation that is perceived with the nose

sneeze

- involuntary expulsion of air from the nose

snow

- frozen precipitation in the form of crystalline ice

soil

- upper layers of Earth in which plants grow

- made of minerals and organic matter

solar cell

- device that uses sunlight to generate electricity

67

solar eclipse

- eclipse of the Sun

- happens when the Moon comes between the Sun and Earth

solar energy

- energy transmitted from the Sun

solar panels shown above are used to capture solar energy

solar system

- Sun and the collection of celestial bodies, including meteors, asteroids, comets, moons, and planets that orbit it

solid

- rigid state of matter that has a definite volume and a definite shape

solute

- substance that dissolves in a solvent

solution

- homogeneous mixture of one or more substances dissolved in another substance

solvent
- substance that is capable of dissolving other substances

sound
- form of energy produced by the vibrations of air molecules

- transmitted in the form of waves

space
- unlimited expanse in which everything is located

space probe
- research spacecraft sent from Earth to study another body in space

species
- group of related organisms that are closely similar in appearance and are capable of interbreeding

spectrum
- complete range of colors or wavelengths

speed
- distance traveled per unit of time

- the rate of motion

speed of light
- speed of light in vacuum

- about 3×10^8 m/s

sperm
- male reproductive cells produced by the testes

spinal cord
- column of nerve tissue that runs through the backbone
- carries most messages between the brain and the rest of the body

spore
- asexual reproductive cell of nonflowering plants, such as ferns and fungi

stamen
- male reproductive structure of a flower

anther

stamen

filament

star
- massive celestial body in outer space consisting of intensely hot gases, such as hydrogen and helium

state of matter
- physical form of matter, such as liquid, solid, or gas

static
- electrical disturbance caused due to atmospheric conditions

static electricity
electrical charge created due to friction between objects

stem
- main axis of a plant bearing the buds, leaves, and flowers
- holds a plant up

stigma
- female part of a flower that receives pollen grains for fertilization

stigma

stomach
- body organ used to digest food

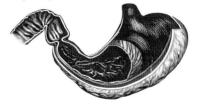

stoma

- pore on a plant's leaves used for respiration

style

- part of flower found between the ovary and the stigma

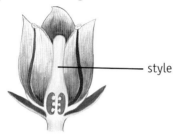

style

submarine

- specialized watercraft that can submerge and operate underwater

subsoil

- layer of soil lying below the surface soil

- made of sand, silt, and clay but little humus or other organic matter

substance

- chemical element, compound, or mixture

sun

- star that is at the center of our solar system

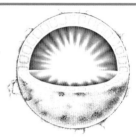

survive

- continue to live

sweat

- salty fluid secreted by sweat glands

switch

- device used for turning an electric circuit on and off

symbol

- letter or letters used to denote a chemical element

tail
- structure at the back end of an animal

tail

taste
- sense that allows one to distinguish the flavors of food

technology
- application of scientific knowledge and processes to solve problems or to make useful tools and other products

telegraph
- machine used for transmitting and receiving messages over long distances using Morse code

telephone
- device that transmits speech by means of electric signals

telescope
- astronomical instrument that enlarges images of distant objects

temperature
- degree of hotness or coldness that can be measured using a thermometer

tendon
- soft tissue connecting a muscle to a bone

terrarium

- glass or plastic container used for keeping and observing small animals or plants

theory

- general principle that explains or predicts facts or events

thermal

- dealing with heat

thermometer

- instrument used to measure temperature

thrust

- force that produces motion

- pushes an airplane forward

thunderstorm

- storm with heavy rainfall, thunder, and lightning

tide

- periodic rise and fall of the ocean's surface due to gravitational forces of the Moon and the Sun

tissue

- group of similar cells, specialized to perform a particular function

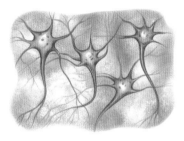

tongue

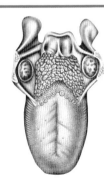

- organ found in the mouth that manipulates food for chewing and swallowing

- organ containing taste buds

tooth

- hard structure found in the jaws of many vertebrates used to tear and chew food

- humans have 20 "baby" teeth and 32 permanent teeth

topsoil
- upper layer of soil in which plants have most of their roots

tornado
- violent, whirling wind characterized by a twisting, funnel-shaped cloud

touching
- detecting one's surroundings through pressure receptors in the skin

trachea
- largest breathing tube in the body that leads from the throat to the lungs

- also called the windpipe

trait
- a genetically inherited characteristic of an organism

transformer
- device used to increase or decrease voltage of electric current

translucent
- allowing some light to pass through

transparent
- allowing light to pass through; able to be seen through

  clear glass is a transparent material

transportation
- means and methods of the movement of persons or goods

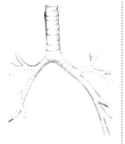

tropical
- zone between the Tropic of Cancer and the Tropic of Capricorn characterized by high temperature, humidity, and rainfall

tropism
- involuntary movement of an organism activated by an external stimulus

heliotropism, movement of plants toward the Sun, is an example of tropism

tsunami
- huge ocean wave produced by an underwater earthquake, landslide, or volcanic eruption

tundra
- arctic and subarctic ecosystem dominated by lichens, mosses, grasses, and woody plants

typhoon
- violent tropical storm or cyclone with very strong winds and heavy rainfall

ultraviolet light
- light rays having slightly shorter
  wavelengths than visible light

universe
- all matter and energy, including
  our solar system, thousands of
  galaxies, and everything else that
  exists

vaccine

- substance introduced into the body to produce immunity against a specific disease

- substance composed of killed or weakened cells, or of proteins that stimulates the production of antibodies

vacuum

- space with very little or no matter

- empty space

vapor

- gaseous state of liquids or solids

variable

- symbol used to represent a value or set of values

- quantity represented by a symbol that can have different values

in $3x + y = 23$, x and y are variables

vegetable

- part of a plant that is commonly consumed as food by humans

carrots, potatoes, cabbages, and beans are examples of vegetables

vein

- blood vessel that carries blood from the body back to the heart

velocity

- the speed of an object moving in a certain direction

vertebrate

- animal that has a backbone

  mammals, birds, reptiles, amphibians, and fishes are examples of vertebrates

vibration

- back and forth movement of an object

- an oscillatory motion

virus

- microorganism that infects cells and causes disease

- organism that invades and grows in cells and thereby alters their ability to function

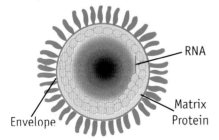

RNA

Matrix Protein

Envelope

vitamin

- organic substance required by the body in very small amounts for healthy growth and development

  vitamins A, C, and E are examples of vitamins

volcano

- vent in Earth's crust through which molten rock, steam, and ash reach the surface

volt

- basic unit of voltage

voltage

- force that pushes electricity through a wire

- the difference in electrical energy between two points in an electric circuit

volume

- amount of space occupied by a three-dimensional object

- measured in cubic units

warm-blooded

- maintaining a warm body temperature, independent of the temperature of the surroundings

mammals and birds are warm-blooded

warm front

- zone where a mass of warm air replaces a mass of cold air

water

- colorless, tasteless, and odorless liquid

- made of H_2O molecules

water cycle

- recycling of water between Earth and the atmosphere

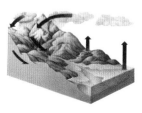

water vapor

- water in a gaseous state

- colorless, odorless, invisible, gaseous form of water in the atmosphere

watt

- basic unit of power

wave

- an up-and-down vibration traveling through water

- any kind of vibration traveling through something

wavelength

- the distance between two wave crests

- denoted by the Greek letter *lambda* (λ)

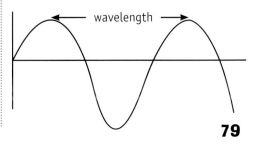

wavelength

weather

- the day-to-day changes in the atmosphere

- specific condition of the atmosphere measured with respect to wind, temperature, air pressure, etc.

weathering

- process of decomposition or disintegration of rocks

- weathering can be caused by physical or chemical processes

web

- intricate network formed by weaving or interweaving

- shelter or food-catching trap made by spiders and other insects

- synonym for World Wide Web

wedge

- an inclined plane that moves

- anything that splits, cuts, or divides another object

  a knife, an axe, and a razor blade are examples of wedges

weight

- the force exerted on an object by gravity

- measured in newtons or pounds

wetland

- area that is partially covered with water throughout the year

marshes, swamps, and bogs are examples of wetlands

whale

- a marine mammal with a body shaped like a dolphin

- the blue whale is the largest mammal, largest vertebrate, and largest known animal

wheel and axle
- a simple machine consisting of a circular frame attached to an axle revolving around a central axis

white blood cell
- type of blood cell that helps the body to fight infections and diseases

white light
- combination of all colors of light
- sunlight is white light

wind
- movement of air from an area of high pressure to an area of low pressure

wing
- modified front limbs of a bird that the bird may use to fly
- any structure on an animal that the animal uses to fly
- large horizontal surface attached to an airplane, which produces lift and allows it to fly

wood
- hard, fibrous organic substance under the bark of trees

work
- the product of the force on an object and the distance the force moves the object
- measured in joules

81

x-ray

- form of electromagnetic radiation of very short wavelength

- rays that can penetrate solid objects

- used in medical practices to make images of bones in the body

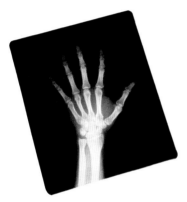

xylem

- tissue found in plants that conducts water from the roots to other parts of the plant

yeast

- group of single-celled fungi

- used in making bread and cheese

zoology

- branch of science that deals with the structure, growth, and classification of animal life

Abbreviations

mm	millimeter		g	gram
cm	centimeter		kg	kilogram
m	meter		t	metric ton
km	kilometer			
			oz	ounce
mm^2	square millimeter		lb	pound
cm^2	square centimeter		T	ton
m^2	square meter			
km^2	square kilometer		mL	milliliter
			L	liter
cm^3	cubic centimeter		kL	kiloliter
m^3	cubic meter			
			fl oz	fluid ounce
in.	inch		c.	cup
ft	foot		pt	pint
yd	yard		qt	quart
mi	mile		gal	gallon
$in.^2$	square inch		° C	degrees Celsius
ft^2	square foot		° F	degrees Fahrenheit
yd^2	square yard			
			A.M.	anti meridiem (morning)
$in.^3$	cubic inch		P.M.	post meridiem (afternoon or evening)

Measurement

Length

10 mm = 1 cm
100 cm = 1 m
1,000 m = 1 km
12 in. = 1 ft
3 ft = 1 yd
1,760 yd = 1 mi

Mass

1,000 g = 1 kg
1,000 kg = 1 t
16 oz = 1 lb
2,000 lb = 1 T

Capacity

1,000 mL = 1 L
1,000 L = 1 kL
8 fl oz = 1 c.
2 c. = 1 pt
2 pt = 1 qt
4 qt = 1 gal

Area

100 mm^2 = 1 cm^2
10,000 cm^2 = 1 m^2

Time

60 seconds = 1 minute
60 minutes = 1 hour
24 hours = 1 day
7 days = 1 week
365 days = 1 year
366 days = 1 leap year
12 months = 1 year
10 years = 1 decade
100 years = 1 century
1,000 years = 1 millennium

Days in a month

30 days   September, April, June, November

31 days   January, March, May, July, August, October, December

28 days   February (most years)

29 days   February (leap year)

Conversions

Length
Standard to Metric

1 in.	2.54 cm
1 ft	30.48 cm
1 yd	0.91 m
1 mi	1.61 km

Metric to Standard

1 cm	0.39 in.
1 m	1.09 yd
1 km	0.62 mi

Capacity
Standard to Metric

1 fl oz	28.41 mL
1 pt	0.57 L
1 gal	4.55 L

Metric to Standard

1 L	1.76 pt

Mass
Standard to Metric

1 oz	28.35 g
1 lb	0.45 kg
1 T	1.02 t

Metric to Standard

1 g	0.04 oz
1 kg	2.20 lb
1 t	0.98 T

Temperature
Fahrenheit to Celsius

$$°C = \frac{5}{9}\,(°F - 32)$$

Celsius to Fahrenheit

$$°F = \frac{9}{5}\,°C + 32$$

The Elements

Name	Symbol	Atomic Number	Name	Symbol	Atomic Number
Actinium	Ac	89	Gallium	Ga	31
Aluminum	Al	13	Germanium	Ge	32
Americium	Am	95	Gold	Au	79
Antimony	Sb	51	Hafnium	Hf	72
Argon	Ar	18	Hassium	Hs	108
Arsenic	As	33	Helium	He	2
Astatine	At	85	Holmium	Ho	67
Barium	Ba	56	Hydrogen	H	1
Berkelium	Bk	97	Indium	In	49
Beryllium	Be	4	Iodine	I	53
Bismuth	Bi	83	Iridium	Ir	77
Bohrium	Bh	107	Iron	Fe	26
Boron	B	5	Krypton	Kr	36
Bromine	Br	35	Lanthanum	La	57
Cadmium	Cd	48	Lawrencium	Lr	103
Calcium	Ca	20	Lead	Pb	82
Californium	Cf	98	Lithium	Li	3
Carbon	C	6	Lutetium	Lu	71
Cerium	Ce	58	Magnesium	Mg	12
Cesium	Cs	55	Manganese	Mn	25
Chlorine	Cl	17	Meitnerium	Mt	109
Chromium	Cr	24	Mendelevium	Md	101
Cobalt	Co	27	Mercury	Hg	80
Copper	Cu	29	Molybdenum	Mo	42
Curium	Cm	96	Neodymium	Nd	60
Darmstadtium	Ds	110	Neon	Ne	10
Dubnium	Db	105	Neptunium	Np	93
Dysprosium	Dy	66	Nickel	Ni	28
Einsteinium	Es	99	Niobium	Nb	41
Erbium	Er	68	Nitrogen	N	7
Europium	Eu	63	Nobelium	No	102
Fermium	Fm	100	Osmium	Os	76
Fluorine	F	9	Oxygen	O	8
Francium	Fr	87	Palladium	Pd	46
Gadolinium	Gd	64	Phosphorus	P	15

The Elements

Name	Symbol	Atomic Number	Name	Symbol	Atomic Number
Platinum	Pt	78	Sodium	Na	11
Plutonium	Pu	94	Strontium	Sr	38
Polonium	Po	84	Sulfur	S	16
Potassium	K	19	Tantalum	Ta	73
Praseodymium	Pr	59	Technetium	Tc	43
Promethium	Pm	61	Tellurium	Te	52
Protactinium	Pa	91	Terbium	Tb	65
Radium	Ra	88	Thallium	Tl	81
Radon	Rn	86	Thorium	Th	90
Rhenium	Re	75	Thulium	Tm	69
Rhodium	Rh	45	Tin	Sn	50
Roentgenium	Rg	111	Titanium	Ti	22
Rubidium	Rb	37	Tungsten	W	74
Ruthenium	Ru	44	Uranium	U	92
Rutherfordium	Rf	104	Vanadium	V	23
Samarium	Sm	62	Xenon	Xe	54
Scandium	Sc	21	Ytterbium	Yb	70
Seaborgium	Sg	106	Yttrium	Y	39
Selenium	Se	34	Zinc	Zn	30
Silicon	Si	14	Zirconium	Zr	40
Silver	Ag	47			

The Periodic Table of the Elements

Key

6
C
Carbon

atomic number — 6
chemical symbol — C
element name — Carbon

1																		2
H																		He
Hydrogen																		Helium

3	4											5	6	7	8	9	10
Li	Be											B	C	N	O	F	Ne
Lithium	Beryllium											Boron	Carbon	Nitrogen	Oxygen	Fluorine	Neon

11	12											13	14	15	16	17	18
Na	Mg											Al	Si	P	S	Cl	Ar
Sodium	Magnesium											Aluminum	Silicon	Phosphorus	Sulfur	Chlorine	Argon

19	20	21	22	23	24	25	26	27	28	29	30	31	32	33	34	35	36
K	Ca	Sc	Ti	V	Cr	Mn	Fe	Co	Ni	Cu	Zn	Ga	Ge	As	Se	Br	Kr
Potassium	Calcium	Scandium	Titanium	Vanadium	Chromium	Manganese	Iron	Cobalt	Nickel	Copper	Zinc	Gallium	Germanium	Arsenic	Selenium	Bromine	Krypton

37	38	39	40	41	42	43	44	45	46	47	48	49	50	51	52	53	54
Rb	Sr	Y	Zr	Nb	Mo	Tc	Ru	Rh	Pd	Ag	Cd	In	Sn	Sb	Te	I	Xe
Rubidium	Strontium	Yttrium	Zirconium	Niobium	Molybdenum	Technetium	Ruthenium	Rhodium	Palladium	Silver	Cadmium	Indium	Tin	Antimony	Tellurium	Iodine	Xenon

55	56	57	72	73	74	75	76	77	78	79	80	81	82	83	84	85	86
Cs	Ba	La	Hf	Ta	W	Re	Os	Ir	Pt	Au	Hg	Tl	Pb	Bi	Po	At	Rn
Cesium	Barium	Lanthanum	Hafnium	Tantalum	Tungsten	Rhenium	Osmium	Iridium	Platinum	Gold	Mercury	Thallium	Lead	Bismuth	Polonium	Astatine	Radon

87	88	89	104	105	106	107	108	109	110	111
Fr	Ra	Ac	Rf	Db	Sg	Bh	Hs	Mt	Ds	Rg
Francium	Radium	Actinium	Rutherfordium	Dubnium	Seaborgium	Bohrium	Hassium	Meitnerium	Darmstadtium	Roentgenium

58	59	60	61	62	63	64	65	66	67	68	69	70	71
Ce	Pr	Nd	Pm	Sm	Eu	Gd	Tb	Dy	Ho	Er	Tm	Yb	Lu
Cerium	Praseodymium	Neodymium	Promethium	Samarium	Europium	Gadolinium	Terbium	Dysprosium	Holmium	Erbium	Thulium	Ytterbium	Lutetium

90	91	92	93	94	95	96	97	98	99	100	101	102	103
Th	Pa	U	Np	Pu	Am	Cm	Bk	Cf	Es	Fm	Md	No	Lr
Thorium	Protactinium	Uranium	Neptunium	Plutonium	Americium	Curium	Berkelium	Californium	Einsteinium	Fermium	Mendelevium	Nobelium	Lawrencium